I0062612

UNBLOCKED
The Power of Real Intelligence in the Era of AI

Copyright © 2025 by **Asim Kumar**

Published by AVIGNA, LLC

https://www.avignallc.com

First Edition

Paperback ISBN: 979-8-9941054-8-1
Ebook ISBN: 979-8-9941054-0-5

Contents

To my wife Sangeeta for being my constant source of Real Intelligence.

To my children, Arul and Adya, for being courageous to find opportunities to create unlimited success.

PREFACE

Thirty Years in the Technology Trenches

I wrote this book because, after three decades in the information technology industry, I realized that the biggest problems businesses face are rarely about the technology itself. The servers usually work. The code usually compiles. The cloud usually scales.

The real problem is human. The real obstacles are fear, complexity, and disconnected thinking.

I have spent the last 30+ years helping organizations navigate these obstacles. But my journey didn't start in a corporate conference room or a high-end data center. It started in my apartment living room.

The Hunger for Knowledge

My career began in the early nineties with a basic need for survival in a tight job market along with simple, burning curiosity. As part of my undergraduate studies, I had changed several majors, ranging from Computer Engineering to Computer Information Systems. One class that really resonated with me was "Data Communications". It made practical sense to me, and I could see myself delving deeper into it. I did some research and figured out that I didn't just want to read about networking; I wanted to get hands-on with it. I wanted to see how it all worked to be more employable. Most importantly, this was my way out of being a software programmer/engineer.

I remember scraping together enough hardware to build a two-computer network in my apartment just so I could install a 5-user version of Novell NetWare. I spent sleepless nights troubleshooting connection issues, figuring out IRQ conflicts, and staring at command lines until my eyes blurred. I didn't stop until I earned my very first certification: Certified Novell Administrator (CNA).

That humble home lab was the seed. It taught me that knowledge isn't given; it is obtained through the friction of doing.

I worked continuously on adding advanced credentials from Novell and Microsoft as my career accelerated. In the latter part of the nineties, I moved into the world of Cisco, becoming a CCIE in the year 2000. I spent years in the trenches of internetworking, consulting, training, project management, and business development. I worked with diverse customers—from Internet Service Providers and conglomerates of retail giants to healthcare systems and government agencies. I saw the evolution of Data Networks and Telecommunications industry firsthand, from the early days of Local Area and Wide Area Networks (LAN/WAN) to the complex, hyper-connected cloud architectures of today.

The Financial Pivot (and the Lesson on Risk and Deep Dive into Human Psychology)

In 2005, after a decade of technical consulting and training and going through the dot.com downturn, I took a risky career detour that surprised many of my peers. The period of weekly layoffs around me had an indirect effect on my psychology, and I started questioning the validity of all the credentials and experience I had earned in networking technology. In my insecurity in the technology arena, I felt an urgent need to diversify and developed a serious interest in the financial markets, illusioned with the better-faster road to personal security and financial freedom.

I didn't just dabble; I dove in without knowing how to swim (I still don't swim). I learned to trade a variety of financial securities (Stocks and Forex) and derivatives (Futures and Options). Over time, I discovered the importance of understanding mindset and self-studied human behavior psychology, and the importance of risk management. In 2007, while working full-time, I even enrolled in a 2 year full-time MBA for

Executives program, and, in the middle of the program, I even started planning for achieving the CFA (Chartered Financial Analyst) designation. At the time, I desperately wanted to switch careers as I just wanted to leave what I was doing in technology and get into finance. The grass was greener on that side in my mind at the time.

Then came the crash of 2008 right in the middle of the MBA program. Watching the markets collapse and the uncertainty that abounds in the field taught me an expensive but invaluable life lesson: managing your energy, Patience, Discipline and Focus (ePDF) is the difference between building wealth and financial drainage. It is now my daily practice to read my own ePDF.

I realized that the "Fear" driving stock market crashes was the same "Fear" I saw not too long ago during the dot.com hype and eventual crash.

- Traders panic-sell when the market drops.

- CTOs panic-buy technology when a security threat looms.

- Investors chase "hype" stocks just like businesses chase "shiny object" technologies.

With my new business credential, I remained in the technology world with a new set of eyes and a better mindset. I wasn't just a highly credentialed network engineer anymore; I was a strategic thinker. I understood that Real Intelligence wasn't just about knowing how to design and implement complex networks; it was about understanding the organizational workings and discipline to stay calm, rational, and strategic when everyone else was freaking out. I broadened my professional scope into Information Security- another field where paranoia is prevalent.

My experience in financial markets and information security gave me a solid foundation in Risk Management and Human Psychology. I was so intrigued by the field of human psychology that I used to joke about my next academic venture being a PhD in human behavior psychology. In my research of psychological studies, I came across NLP (Neuro-Linguistic Programming) and I became certified as a Master Practitioner to test the newly found waters, again, knowing that I do not swim. Interesting enough, we use a different form of NLP (Natural Language Processing) in today's Artificial Intelligence world of technology. So, here I am now, writing this book, connecting the dots between the two forms.

I want to mention "Acres of Diamonds", a famous story by Russell H. Conwell about a Persian farmer who becomes discontented after hearing about diamonds, sells his prosperous farm, and travels the world fruitlessly searching for them, only to die poor, while the new owner of his land discovers the legendary "diamond mines of Golconda" in his own backyard, illustrating that opportunities for wealth and success are often right where you are, if you have the vision to see them. I have written this book going through my journey of finding my "Acres of Diamonds" and believe this will help you find yours.

The Mission: Removing Obstacles

In 2013, I founded my own consulting firm, **Avigna**, pronounced (**a-wig-ná**). In Sanskrit, the word *Avigna* means "**Remover of Obstacles**."

This is not just a business name; it is my philosophy. Whether I am writing a multi-million dollar proposal or helping an organization modernize its technology, migrate to the cloud, my goal is to remove the blockages that stop success.

This book is the culmination of those 30+ years. It is a synthesis of the hard skills I learned to build networks and the mindset skills I learned in the financial markets.

We are living in an era of noise. Today, Artificial Intelligence is reshaping our world, and it is creating a new wave of fear and confusion. I wrote this book to show you that you don't need to be afraid. You just need an actionable methodology.

If you can combine the speed of Artificial Intelligence with the wisdom of Real Intelligence,

you won't just survive the changes coming our way.
You will be **Unblocked**.

A NOTE TO THE READER

Who is this book for?

This book is written in the language of business. I use terms like *Product*, *Service*, and *Profit*.

But do not be mistaken. **This is not just a book for corporations.**

- **If you are a CEO or Tech leader,** you will use these frameworks to fix your company culture, optimize your technology and cloud spend, and align your teams.

- **If you are a Solopreneur, Freelancer, or an Individual,** you are the CEO of *"You Inc."* You will use these frameworks to fix your mindset, optimize your habits, and align your career.

The laws of physics are the same whether you are building a bridge or building a life. As you read, whenever you see the word "Company," I invite you to substitute it with "My Life." Whenever you see "Technology," substitute it with "My Tools."

The obstacle is universal. And so is the path to becoming **Unblocked**. To be unblocked is to replace fear-based reaction with value-based action, allowing you and your teams to operate in a state of alignment and flow.

CHAPTER 1: INTRODUCTION

The Solutions Methodology

Every once in a while, a shift happens that changes everything.

We saw it with the Internet in the 90s. We saw it with the dotcom in the 2000s. We saw it with the cloud in the 2010s. And now, we are seeing it with Artificial Intelligence.

But unlike previous shifts, this one isn't just changing ***what*** we use; it is changing ***how*** we think. And for many leaders, it is terrifying.

In times of massive shifts, the natural human reaction is to panic. We freeze. We hoard. We react. I call this the ***Era of Noise***.

Whether you are a CEO looking at a balance sheet or an IT Director looking at a system dashboard or server log, you are bombarded with it.

- *"Buy this Generative AI tool, or you'll be obsolete!"*

- *"Secure this cloud environment, or you'll be hacked!"*

- *"Cut this cost, or you'll miss quarterly targets!"*

The noise creates Fear. And Fear is the enemy of Intelligence. When we operate out of fear, our vision narrows. We stop looking for opportunities and start looking for threats. We stop building and start protecting ourselves.

I wrote this book to cut through the noise and help you start thinking with a different mindset. To do that, I want to answer four simple questions: Why did I write this book? When does this matter? How will it help you? And What exactly is the solution?

1. THE WHY

Why did I write this book?

I wrote this book because I am tired of seeing the disconnect between Business and Technology.

In my 32 years of consulting, I have walked into a variety of companies—from Fortune 500 giants to scrappy startups. And I see the same tragedy playing out over and over again.

I see Business leaders demanding speed without understanding the risk.

I see Technology leaders demanding budget without explaining the value.

They are working in the same building (or on the same Zoom call), but they are living in different realities. They speak different languages. And because of that disconnect, they operate out of Fear.

- Business fears falling behind. They read headlines about competitors using AI to double productivity, so they put pressure

on IT to "do something with AI" by next
week.

- The Tech Team fears being blamed. They
  know the infrastructure isn't ready. They
  know the security risks. So, they say "No,"
  or they build shadow systems that don't
  scale.

This disconnect creates a blockage to the business's
growth. Ideas get blocked by bureaucracy.
Innovation gets blocked by legacy debt. Talent gets
blocked by a bad culture.

I have had my experiences in all these environments,
and I was able to get around by venturing
into unfamiliar territories of finance and human
psychology with a belief as elegantly quoted by
Eleanor Roosevelt: *"**You gain strength, courage,
and confidence by every experience in which
you stop to look fear in the face...You must do
that which you think you cannot do.**"*

This is one of my favorite quotes that I often put
into practice for myself in uncertain and difficult

situations. I wrote this book to help you do the thing which you think you cannot do: bridge the gap between Business and Technology. Bridge the gap between "You" and Technology.

I believe that Real Intelligence (RI)—your empathy, your judgement, your wisdom—is the only thing that can align these two worlds.

The "Villain" of this Story: **The Reactive Leader**

Throughout this book, we will fight against a specific character type: the Reactive Leader. The Reactive Leader is smart, hardworking, and well-intentioned. But they are dangerous. This behavior can be due to their emotional setup as part of their personal upbringing and experiences.

- They buy software because a competitor bought it (**FOMO** – Fear Of Missing Out).

- They hire people based on keywords in a resume, not character.

- They slashed training budgets to save money in Q3, only to pay 10x that amount in outages in Q4.

The Reactive Leader is blocked. This book is the manual for becoming a Proactive Solution Architect of your business and life.

2. THE WHEN

Why does this matter NOW?

You might ask, *"Asim, can't I read this next year? We are really busy right now."*

My answer is simple: The margin for error has disappeared.

Ten years ago, if IT and Business weren't aligned, you lost some efficiency. Maybe you wasted 10% of your budget. It hurt, but it wasn't fatal. Today, the stakes are existential.

- **Security:** If you ignore security today, you don't just lose data; you lose your reputation. In the trust economy, once reputation is gone, the business is dead.

- **Artificial Intelligence**: If you ignore AI,

you don't just fall behind; you become obsolete. The speed of innovation has moved from linear to exponential.

- **The Human Element:** If you ignore your people, you burn out your workforce in a month. The modern employee will not tolerate "grind culture" for a paycheck. They will leave, taking their tribal knowledge with them.

The Future Look: 2030 and Beyond

We are entering a world where "Technical Skill" will be a commodity. AI will write code (it is already doing it if you look at the right places). AI will configure servers. AI will detect threats. So, what is left for us?

Real Intelligence. The leaders who thrive in the next decade won't be the ones who know the most command-line codes. They will be the ones who can align People, Process, and Technology. The time to modernize your **mindset** is not "someday." It is Now.

3. THE HOW

How will this book help you?

This is not a textbook. Textbooks are for memorizing; this book is a manual for executing.

I am not interested in giving you a theory that you put on a shelf and ignore. I am interested in Action. I have designed this book as a Solutions Methodology.

It works by removing obstacles. (That is the meaning of my company name, **Avigna**). We will do this by installing a new operating system in your leadership style using three core phases:

PART 1: We will Align (The Foundation):

We will balance the tripod of People, Process, and Technology (PPT). Most companies focus on Tech first. We will flip that script

PART 2: We will Measure (The Metrics):

We will learn how to select a P-R-O-D-U-C-T based on integrity, not hype.

We will deliver S-E-R-V-I-C-E that builds partnerships, not just tickets.

We will drive P-R-O-F-I-T that fuels sustainability, not greed.

PART 3: We will Modernize (The Execution):

We will use the E.A.S.I.E.R. cloud strategy to make technology serve the business.

We will redefine W.O.R.K. not as a grind, but as the pursuit of knowledge.

We will use F.O.C.U.S. to direct our energy toward unlimited success.

This methodology helps you by giving you the words, the logic, and the confidence to say "No" to the noise and "Yes" to the signal.

4. THE WHAT

So, what is this book about?

It is a manifesto for **Real Intelligence (RI)**.

It is a guide that argues:

- Technology is just an amplifier. You are the source code.

- Profit is not greed; it is Oxygen for you and your business.

- Work is not a grind; it is your path to Knowledge.

I am presenting you with a choice. You can continue to operate in the old world—siloed, reactive, and fearful. You can stay blocked. Or, you can turn the page, and we can start building a future that is Elastic, Agile, and Unlimited.

Welcome to the Avigna Methodology.

THE AVIGNA DECISION MATRIX: Are You Blocked?

(An interactive audit to gauge your current state)

Before we begin the work, we must assess the current reality. Be honest with yourself as you answer these three questions.

1. The "**Why**" Audit

- *Question:* When you or your organization makes a major technology purchase, what is the primary driver?

- *Blocked Answer:* "Because we are afraid of falling behind" or "Because our competitor did it." (Fear).

- *Unblocked Answer:* "Because it solves a specific process bottleneck that helps our people serve customers better." (Real Intelligence).

2. The "**Who**" Audit

- *Question:* If your most "technical" employee left tomorrow, would your business grind to a halt?

- *Blocked Answer:* "Yes, Bob knows everything. If he leaves, we are in trouble." (Tribal Knowledge).

- *Unblocked Answer:* "No, because we have

documented Processes and a culture of shared knowledge." (Resilience).

3. The "**How**" Audit

- *Question:* How do you or your team react to a crisis (e.g., a system outage)?

- *Blocked Answer:* Finger-pointing, panic, and silence. (Fear).

- *Unblocked Answer:* Collaboration, root-cause analysis, and communication. (Trust).

If you found yourself leaning toward the "Blocked" answers, this book was written for you. Let's get to work.

PART 1

Let's Align

CHAPTER 2: THE FOUNDATION

The Tripod of Success

If you walk into any corporate meeting in the world and ask why a strategic initiative failed, you will hear a laundry list of excuses.

"The software was buggy."

"The timeline was too aggressive."

"The budget got cut."

"The vendor didn't deliver."

But in my 30+ years of consulting, I have found that these are rarely the root causes. They are symptoms. They are cracks in the drywall that appear when the foundation is sinking.

The root cause is almost always an imbalance in the foundation.

In our industry, we often talk about **PPT**: **People, Process, and Technology**.

It is a concept that has been around since the 1960s (originally Harold Leavitt's "Diamond Model"), but it is often misunderstood as just a checklist. Leaders treat it as three separate silos. HR handles People. Operations handle Process. IT handles Technology.

The Avigna Methodology views PPT differently. We view it as a **Three-Legged Stool**.

- **Leg 1:** The People (The Heart & Real Intelligence)

- **Leg 2:** The Process (The Map & Direction)

- **Leg 3:** The Technology (The Tools & Amplifier)

The physics of a three-legged stool are unforgiving. If you shorten one leg, the stool doesn't just wobble; it becomes useless.

- If you have the best Technology money can buy, and the most detailed Process documents ever written, but your People are disengaged? **The stool falls over.**

- If you have motivated People and great Technology, but no Process? **You have chaos.**

This chapter is about finding the balance—the "Sweet Spot"—where your business becomes stable, scalable, and **Unblocked.**

1. THE FIRST PILLAR: PEOPLE

The Source of Real Intelligence

Let's start with the most critical, yet most neglected, leg of the stool: **The People.**

In the modern business landscape, we have become obsessed with automation. We track server uptime, we monitor cloud latency, and we measure code deployment velocity. But we rarely measure the "uptime" of our people.

In the age of AI, it is tempting to view people as "overhead"—a line item on a spreadsheet to be minimized. I have heard many conversations where executives proudly announced they were replacing customer service agents with a chatbot to "increase efficiency." Six months later, their Net Promoter Score (NPS) had crashed, and their best clients were leaving. Why? Because they forgot a fundamental truth:

Technology provides speed, but People provide *context*.

Technology provides data, but People provide *wisdom*.

If you treat your people like cogs in a machine, they will break like cheap plastic. But if you treat them as the source of **Real Intelligence (RI)**, they become the strongest asset on your balance sheet.

The "Villain": The Knowledge Hoarder

Every organization has a villain that blocks the People pillar. I call this character **The Knowledge Hoarder** (or the Brilliant Jerk).This is an engineer who knows everything but listens to nothing. They refuse to document their work because "it's all in their heads." They treat knowledge sharing as a personal boundary issue rather than a professional necessity. They make themselves indispensable by creating bottlenecks. These people are not necessarily bad people and they behave in such a way due to their own personal insecurities and habits.

Early in my career, I worked with a network engineer who fit the description of the "Brilliant Jerk" perfectly. He was exceptionally skilled—there was no denying that—but he carried his knowledge like a private asset.

When the systems went down, he was always the hero who "saved the day," but only because he had ensured that no one else could even begin to diagnose the problem. His attitude discouraged junior staff from asking questions. Over time, the entire team began working *around* him instead of *with* him. Productivity didn't collapse overnight, but it slowed subtly and steadily. One Brilliant Jerk, I learned, can quietly drain the energy of ten capable people. This is not one such story. Throughout my consulting career working with several clients, I have had the experience of running into this personality type on numerous occasions.

In contrast, I once hired a young engineer who arrived with very little technical experience but an impressive **Willingness** to learn. He asked thoughtful questions, took notes, and treated every challenge as an opportunity. I still remember him staying late—not because anyone asked him to, but because he genuinely wanted to understand how things worked. Within a couple of years, he had developed into one of the most reliable engineers on the team. Fortunately, I have had the privilege of running into this personality type more often than the previous one. Goodness prevails.

The Avigna Insight: Technical brilliance can be valuable, but a willingness to grow is transformational. It is the difference between becoming a bottleneck and becoming a force multiplier.

The Old Way (Fear-Based) vs. The Avigna Way (Mindful)

Hire for Skills: "Get the guy with the most certs." **vs. Hire for Will**: "Get the person with the most curiosity."

Culture of Silence: Don't report bugs; you might get blamed. **vs. Culture of Safety**: Report bugs early so we can fix them.

Hero Mentality: Rely on one genius to save the day. **vs. Team Mentality**: Rely on shared knowledge to prevent the crisis.

The Culture of Emotional Intelligence (EQ)

Once you have the right people, how do you keep them? The biggest barrier to success in the "People" pillar is rarely competence; it is **Fear**. In many IT organizations, there is a culture of silence.

- A junior developer sees a potential bug in

the code but doesn't speak up because the
Lead Architect is intimidating.

- A project manager knows the timeline is
  impossible, but says "Yes" to the leadership
  because they are afraid of looking weak.

This silence is costly. It results in product launches
that crash on Day 1. It results in security breaches
that could have been prevented. Your job as a leader
is to build a culture of **Emotional Intelligence
(EQ)**. EQ in business isn't about "holding hands"
or "singing Kumbaya." It is about **Psychological
Safety**. It means creating an environment where
the truth travels faster than the lies.

In 2002, I was brought into a client company
that was preparing to deploy a new Voice over
Internet Protocol (IP) (VoIP) based telephony
system. At that time, VoIP still felt like a promising
teenager—full of potential, but not quite mature
enough to handle everything it claimed. The vendor
partner who engaged us knew this all too well. Their
previous consultants had left the client frustrated,
and the relationship was hanging by a thread. By the

time I arrived, the atmosphere felt like everyone was holding their breath.

My account team was under pressure from the vendor, and naturally that pressure rolled downhill. Their message to me was simple: "We need this project to land smoothly." My role, however, was equally simple: tell the truth about what the technology could do today—not someday.

During one early meeting, the client's Vice President pulled me aside. He didn't waste time in small talk. "Will this solution even work the way we want it to work?" he asked. Not "Can you install it?" Not "How long will it take?" Just a direct question from someone who had clearly been over promised before.

I took a breath and chose honesty. "It will work," I told him, "but not perfectly on day one. Some features are still on the roadmap. This technology is evolving, and it needs time to settle in to go beyond the legacy system. But with proper transition planning, it will get there."

He nodded slowly. It wasn't the answer he wanted, but it was the answer he needed. That moment became a turning point. Once he saw that I wasn't there to sell him an illusion, trust began to build. And with trust, the conversations shifted from fear and suspicion to collaboration and problem-solving.

Looking back, that project reminded me of a truth I've carried throughout my career:

You can sell optimism, but you can only earn trust with reality.

The "People" Checklist

Before you move to the Process or Technology pillars, audit your People pillar using this checklist:

1. **The "Bus" Factor:** If your most knowledgeable employee got hit by a bus tomorrow, would your business survive? If the answer is "No," you have a People problem. You are relying on "Heroes" rather than a Team.

2. **The "Why" Test:** Walk up to your lowest-level employee and ask them *why* the company exists. If they say "to make money," you have failed to communicate the vision. If they say, "to remove obstacles for our customers," you have Alignment.

3. **The Ego Check:** Who on your team takes credit for success but blames others for failure? That person is a liability, no matter how talented they are.

2. THE SECOND PILLAR: PROCESS

The Map to Consistency

If People are the engine of your company, **Process** is the steering mechanism. Without it, you have a lot of power but no direction.

In my consulting practice, I often walk into a client's office and ask to see their "Process for X" (e.g., onboarding a user or patching a vulnerability). The most common answer I get is: *"Well, usually Bob handles that. Let me ask Bob."*

This is **Tribal Knowledge**. It is an enemy of scale. If "Bob" leaves the company, the process leaves with him.

The Myth of Bureaucracy

Many leaders fear Process because they confuse it with **Bureaucracy**.

- **Bureaucracy** is a wall. It is designed to slow things down, cover backsides, and create paperwork.

- **True Process** is a highway. It is designed to let you move at maximum speed without crashing.

The Avigna Definition:

Process is the removal of decision fatigue.

Think about Steve Jobs or Mark Zuckerberg wearing the same clothes every day. That was a "Process." It eliminated a low-value decision (what to wear) so they could focus their **Real Intelligence** on high-value decisions. Your business processes should do the same. They should automate the routine so your people can innovate on the complex.

Over the years, if there is one pattern that has repeated itself more times than I can count, it's this: when there is no defined process, people start relying on memory and luck. Eventually, luck gives out.

One of my Telecommunications Service Provider clients stands out. They were struggling with recurring network outages—mostly short term, but frequent enough to cause unpleasant customer interactions. Every time an issue occurred, the root cause was almost always the same: an unreviewed configuration change made in good faith, but without proper documentation of the end-to-end system or a second set of eyes.

As the Subject Matter Expert on the account, I suggested something that sounds complex: an inter-departmental **peer-review step** for every configuration change within each responsible department of the customer change request. Nothing complicated. No new software. Just one engineer from each department reviewing another's plan before any misconfigured change propagates to production. A checklist, a quick conversation, and a documented provisioning record.

There was resistance at first—mainly around "slowing things down." But we got the upper management involved and the new process was implemented. Within a few weeks, the tone shifted. Outages dropped dramatically. Engineers across the board began catching each other's small mistakes before they became large problems. New synergies were established.

The Lesson: Technology may be complex, but improvement often begins with one simple, disciplined process. A checklist never gets tired. And it never forgets.

3. THE THIRD PILLAR: TECHNOLOGY

The Amplifier

Finally, we arrive at the third leg: **Technology**.

This is a shiny object. It is where the budget goes. It is where the excitement is. But in the **Avigna**

Methodology, Technology is the *last* step, not the first.

The Golden Rule: Technology is an **Amplifier**.

- If you apply technology to an efficient process, you get **Efficiency**.

- If you apply technology to a broken process, you get **Chaos at the Speed of Light.**

The "Silver Bullet" Trap

I call this the **"Silver Bullet Myth."** Leaders often buy a new software platform (CRM, ERP, Cloud Migration) hoping it will fix a "People Problem" or a "Process Problem."

- *Problem:* "Our sales team isn't closing deals."

- *Wrong Solution:* "Let's buy Salesforce!"

- *Result:* Now you have a sales team that isn't closing deals, and you are paying $50,000 a year for Salesforce. It is not a Salesforce problem. The reason behind deploying is.

I recall the story of a midwestern furniture company. They had been making office furniture for more than a decade. They had a loyal customer base that appreciated the craftsmanship. Over time, the leadership became convinced that the future was automation. They invested heavily in new machinery, believing that technology would give them greater efficiency.

As the automated tools arrived, the number of craftsmen declined. At first, productivity increased. But they missed something important: while the machines were excellent at repeating instructions, they were terrible at interpreting changing customer preferences.

When customer tastes began shifting, the machines kept producing the same patterns. Meanwhile, the craftsmen who once adapted to subtle trends were no longer there to provide that human judgment. The technology was working perfectly; the business was not.

The Avigna Insight: Technology is most valuable when it enhances people and supports a well-designed process—not when it replaces both.

THE INTERSECTION: The "Sweet Spot" vs. The Danger Zones

When you look at these three circles—People, Process, Technology—the magic happens in the center. But the danger happens in the overlaps.

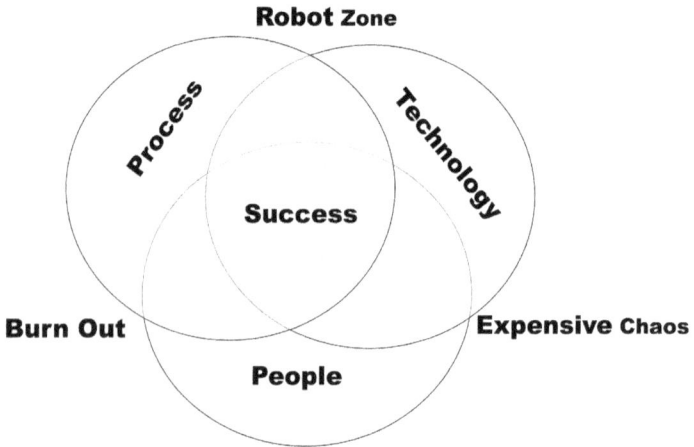

FIGURE: PPT VENN DIAGRAM

Danger Zone 1: People + Tech (No Process) = "Expensive Chaos"

You have smart people and expensive tools, but no rules. Everyone does things their own way. It's the

"Wild West." You burn cash and get inconsistent results because there is no standardization.

Danger Zone 2: People + Process (No Tech) = "Burnout"

This is the "Old School" company. The team is working hard. The rules are clear. But they are doing everything manually. They cannot scale. Eventually, your best talent quits because they are tired of doing robotic work that a computer should be doing.

Danger Zone 3: Process + Tech (No People) = "The Robot Zone"

You have automated everything. On paper, it looks efficient. But when a customer has a unique problem, the system breaks. There is no empathy. There is no **Real Intelligence (RI)** to solve the edge cases.

The Goal: The Sweet Spot = Success

Aligned People, following a Smart Process, amplified by the Right Technology.

FUTURE LOOK: PPT in the Age of AI

How will Artificial Intelligence change this framework in the next 5 years?

You might be thinking, "Asim, if AI is so powerful, won't it eventually replace the People and the Process? Won't we just have 'Technology' left?"

My answer is **No.** In fact, the opposite is true. As technology becomes infinite and cheap, the value of the other two legs—People and Process—further increases. AI will not replace PPT; it will **elevate** it. Here is how the three pillars will evolve:

1. PEOPLE: From Syntax to Strategy

For the last 30 years, we hired technical people for their ability to speak the language of machines. We hired them to know command-line syntax, error codes, and complex configuration scripts. AI has already mastered that language.

- **The Shift:** As AI handles the rote tasks (coding, data entry, syntax checking), your People must become more human.

- **The New Skill Set:** The most valuable engineer in 2030 won't be the one who types the fastest. It will be the one with the highest **Real Intelligence (RI)**: Empathy, Negotiation, Strategic Thinking, and Ethics.

- **The Prediction:** "Soft Skills" will be renamed "Power Skills." You will hire for character and train for AI.

2. PROCESS: From Static to Dynamic

Today, a "Process" is usually a PDF document stored on a SharePoint drive that nobody reads. It is static. It is obsolete the moment it is written. AI will kill the static document.

- **The Shift:** Processes will become dynamic "Guardrails." An AI agent will observe your workflow in real-time. If a team member deviates from the safety protocol, the AI will nudge them back on track instantly.

- **The Result:** We move from "Compliance" (checking a box once a year) to "Continuous Alignment." The process adapts to the situation, rather than constraining it.

3. TECHNOLOGY: From Tool to Partner

Historically, technology was a tool. A hammer doesn't have an opinion. A spreadsheet doesn't give advice. You use it, and you put it away.

- **The Shift:** Tech is moving from "Passive Tool" to "Active Partner."

- **The Scenario:** You won't just "use" a dashboard. You will collaborate with an AI Agent. You will ask it, "What am I missing here?" and it will reply, "You forgot to account for the currency exchange risk in the Asia market."

- **The Danger:** If you treat this Partner like a Tool (ignoring its input), you waste it. If you treat it like a Master (blindly following it), you risk disaster. You must treat it like a **Junior Analyst**—smart, fast, but requiring supervision.

The Avigna Conclusion: The **PPT** framework isn't going away; it is becoming the only competitive advantage left. In a world where every competitor has access to the same "Infinite Technology," the companies that win will not be the ones with the best AI. They will be the ones with the **best Humans** directing that AI.

THE PERSONAL PIVOT: The PPT of "You Inc."

You might be reading this and thinking, *"Asim, I'm a solopreneur,"* or *"I'm just trying to manage my career." How is all this PPT stuff applicable to me?*

Here is the secret: **The laws of physics that govern a billion-dollar corporation are the exact same laws that govern your life.** You are the CEO of "You Inc." And if you feel blocked, overwhelmed, or stuck, it is usually because your personal PPT is out of balance.

1. PEOPLE = Your Mindset & Inner Circle

In a company, "People" are the employees. In your life, "People" is the voice in your head and those five to six people you spend the most time with.

- **The Audit:** Do you have a "Brilliant Jerk" living in your head? Is your inner monologue critical, hoarding fear, and refusing to learn? You need to fire that voice.

- **The Fix:** Hire a mindset of **Willingness**. Surround yourself with people who amplify your Real Intelligence, not your anxiety.

2. PROCESS = Your Habits

The Wisdom: As Aristotle observed, ***"We are what we repeatedly do. Excellence, then, is not an act, but a habit."*** this is the essence of a well-defined personal process—it automates excellence. Your personal process isn't what you say you do; it's what you actually do every morning. A company relies on workflows; you rely on your habits.

- **The Audit:** Do you wake up and react to your phone? That is a broken process. It creates decision fatigue before 9:00 AM.

- **The Fix:** Build a "Personal Kaizen" loop. If a habit drains your energy (like doom-scrolling), engineer it out of your life. Good habits are the roadmap that keeps you on track when motivation fades.

3. TECHNOLOGY = Your Tools

- **The Audit:** Does your phone work for you, or do you work for your phone? Are the apps you use "Amplifiers" of your output, or are they distractions disguised as productivity?

- **The Fix:** Be ruthless. If a tool doesn't serve your mission, delete it.

THE AVIGNA DECISION MATRIX: PPT Audit

Before moving to the next chapter, ask yourself these four questions to audit your foundation.

1. WHY are we struggling?

- Is it a lack of tools? A lack of clarity? Or a lack of skill? (Diagnose the leg of the stool).

2. WHEN do we introduce new tech?

- *Wrong Answer:* "As soon as it comes out."

- *Right Answer:* "Only after we have defined the Process and trained the People."

3. HOW do we capture knowledge?

- If your best employees leave today, do they

take the company's Intellectual Property (IP) with them? If yes, you need to document your Process immediately.

4. WHAT is the culture?

- Do you punish mistakes (Fear), or do you inspect mistakes to improve the Process (Growth)?

Your Thoughts and Self Reflection

(Write your key points and any personal experience related to the ideas presented in the chapter)

PART 2

Let's Measure

CHAPTER 3: PRODUCT

The Selection Methodology

We have established the Foundation (PPT). Now, we must make decisions. One of the most frequent—and stressful—activities in business is **Procurement**: Buying solutions.

In the world of sales and procurement, fear is the dominant currency.

If you are a **Buyer**, you are afraid of making a mistake. You are afraid of buying a solution that fails, crashes the network, or goes over budget. There is an old saying in the IT industry: *"Nobody gets fired for buying IBM."*

That saying is rooted in fear. It means: *I will buy the safe, expensive option not because it is the best fit, but because it protects my job if things go wrong.*

If you are a **Seller**, you are often taught to exploit this fear. I have sat in countless meetings where a salesperson implies, *"If you don't buy our firewall, you will be hacked tomorrow."*

This dynamic—Fear-Based Buying colliding with Fear-Based Selling—leads to a phenomenon I call **Strategic Paralysis**. We either buy nothing (stagnation), or we buy the wrong thing out of panic (waste).

To break this cycle, we need a **Mindful** approach. I developed the **P-R-O-D-U-C-T** framework to give you a rational lens to evaluate any solution. Whether you are selling a solution or buying one, if it meets these seven criteria, you aren't gambling; you are investing.

THE FRAMEWORK: P-R-O-D-U-C-T

P—PERFORM: The Integrity Test

The first letter is the baseline. Does the product do what it says it will do? This sounds obvious, but you would be shocked at how many technology decisions are based on "Roadmap Features"—things the product *might* do in six months—rather than what it can do today.

Performance is not just about speed or specs; it is about **Integrity**. In my consulting work, I often see clients dazzled by a demo that shows 100 bells and whistles. But when we install it, the one critical function they actually need fails.

A few years ago, I helped an enterprise evaluate SD-WAN (Software-Defined Wide Area Network) solutions. The technology offered real promise—better flexibility and centralized management.

After a detailed assessment, the client decided to go with the safest, most established vendor. On paper, it looked like a low-risk decision. In practice, it was anything but. From day one, deployment issues surfaced. Devices wouldn't register properly and connect to the network. Policies applied inconsistently. The vendor blamed early software releases. The client blamed the vendor.

But the deeper problem wasn't just the technology—it was that the client bought the **Marketing**, not the **Performance**. They assumed that because the vendor was "Big," the product was "Ready." It wasn't.

The Avigna Rule: Never buy a promise. Buy performance. If the product cannot solve your problem *today*, it is not a solution; it is a speculation.

R—REVOLUTIONIZE: The Evolution Check

There are two types of products you will buy in your career:

1. **Maintenance Products:** These keep the lights on. They replace "Like for Like." (e.g., buying a faster server to replace an old server).

2. **Evolutionary Products:** These change *how* you work. (e.g., moving from a local file server to a collaborative Cloud platform).

When I evaluate a significant investment, I ask: *Does this evolve the business?* If you are spending significant capital, you should demand significant evolution. Don't just patch the hole; fix the structural weakness.

O—OPTIMIZE: The Friction Test

An optimized product respects your most valuable resource. And no, that resource is not money—it is **Energy**.

Does this product make life easier for your team, or does it add friction? I have seen companies buy "Enterprise Grade" software that is so complex it requires a PhD to operate. It requires 15 clicks to do what used to take two. **The result?** The team finds a workaround. They go back to using Excel spreadsheets because the "Optimized Tool" was too hard to use.

Real Intelligence asks: *Does this tool remove steps, or add them?*

D–DURABLE: The Financial Risk

Coming from my background in financial risk management, I look at every IT purchase as an asset class. Is this product **Durable**?

"Cheap" is often the enemy of Durable. If you buy a cheap solution from a startup that runs out of cash in six months, your investment goes to zero. You have to buy the solution twice. That is the most expensive way to do business.

Durability Checklist:

- **Vendor Stability:** Will this company still

be around in 3 years?

- **Scalability:** Will this product still work if we double our staff next year?

U−USABLE: The Empathy Test

This connects back to the **People** pillar. You can buy the most powerful software in the world, but if the User Interface (UI) is confusing, your team will not use it. Period.

I call this the **Empathy Test**. Did the creators of this product understand the human who has to use it?

- **The Fear-Based Decision:** Buying the tool the CIO likes because it has great reports.

- **The Mindful Decision:** Buying the tool the *employees* like because it helps them do their job.

If adoption is low, the Return On Investment (ROI) is zero. Usability is not a "nice to have"; it is a financial necessity.

C–CONSISTENT: The Trust Factor

In business, we pay for boredom. We want our electricity to be boring (always on). We want our payroll to be boring (always accurate).

Consistency is the bedrock of trust. If a product works perfectly on Monday but crashes on Tuesday, it creates anxiety in your team. In my experience in the financial markets, **Volatility** is where you can make money (if you know what you are doing). But in IT infrastructure, Volatility is where you lose your job. Look for the boring, consistent track record.

T–TRANSPARENT: The Clarity Check

Finally, is the product **Transparent**? This applies to both the technology and the business model.

- **The Technology:** Can we see what the code is doing? Do we have access to the logs? Or is it a "Black Box" that we have to trust blindly?

- **The Business Model:** Are there hidden fees? Surprise licensing costs? "Gotcha" clauses in the renewal?

The Avigna Insight: Hidden costs are obstacles.

- **Advice to Sellers:** Be brutally honest about what your product *cannot* do. You will build more trust by admitting a limitation than by faking a capability.

- **Advice to Buyers:** If the vendor cannot explain the pricing model on a napkin, walk away.

THE IMPLEMENTATION: The Value A.D.D

Selecting the right **P-R-O-D-U-C-T** is only half the battle. You can buy the perfect tool, but if you drop it into a chaotic environment, it will fail. This is where many consultants walk away. They sell you the widget and wish you luck.

But to truly succeed, you must apply the **Value A.D.D** methodology. This is the bridge between "Buying" and "Using."

1. ASSESS (The Audit)

Before you buy, you must understand the *Why*. We start with a "Current State Assessment." We look at your People, your Processes, and your existing Technology. We find the gaps.

- *The Trap:* Prescribing medication before diagnosing the patient. Never buy a solution until you have fully assessed the problem.

2. DESIGN (The Blueprint)

Once we know the problem and have selected the Product, we design the integration. How does this new tool fit into your existing workflow? Who needs to be trained? What processes need to change? Design is where we align the **PPT** (People, Process, Technology). We map it out before we spend a dime on implementation.

3. DELIVER (The Rollout)

This is the execution. Delivery is not just "installing the software." It is **ensuring adoption**. It involves training the humans (**People**), updating the handbooks (**Process**), and configuring the settings (**Technology**).

FUTURE LOOK: Procurement in the Age of AI

How will Artificial Intelligence change how we buy products?

For the last 40 years, IT procurement has been a game of "Liar's Poker." A vendor sends you a brochure claiming their product is the fastest, most secure, and most scalable. You send them an RFP (Request for Proposal) with 500 questions. They answer "Yes" to every single question. You buy the product, install it, and realize... it doesn't work in your environment.

We are about to see the death of the RFP and the rise of **Algorithmic Procurement**.

The Rise of the "Digital Twin"

In the near future, we won't just look at feature sheets. We will use AI Agents to test products for us in a virtual sandbox. Before you spend a dime, your AI will create a "Digital Twin" of your network—a perfect virtual replica of your traffic, your security policies, and your users.

Then, it will invite the Vendor's AI to deploy their product into that Digital Twin.

- **The Prediction:** Your AI will run a thousand simulations overnight. It will tell you: "The vendor claims 99.99% uptime. However, in our specific architecture, when traffic spikes on 'Cyber Monday,' this product creates a bottleneck that will crash the checkout page."

From "Buyer Beware" to "Seller Beware"

This shifts the power dynamic completely. Vendors will no longer be able to hide behind marketing hype or "Roadmap Promises." **Transparency** (The 'T' in our framework) will be forced upon them.

If the code is bad, the AI will find it before the contract is signed.

The Role of Real Intelligence: The Strategic Fit

So, if AI does the testing, what do **You** do? You make the judgment call on **value** and **culture**. AI can tell you if the software works. It cannot tell you if the Vendor is a good partner.

- AI says: "Product A is 5% faster than Product B."

- Real Intelligence says: "Yes, but Vendor B has a dedicated support team in our time zone and a history of ethical behavior. We are going with Vendor B."

The Avigna Conclusion: Procurement will become faster and more honest. But the final signature will still require the gut check of a human who understands the business vision, not just the technical specs.

THE PERSONAL PIVOT: You Are The Product

In the gig economy and the age of AI, the most important product you will ever sell is **Yourself**. Whether you are an employee hoping for a promotion or an entrepreneur pitching investors, you must view yourself through the **P-R-O-D-U-C-T** lens.

- **PERFORM:** Do you deliver? Integrity isn't just for software; it's for people. Do you do what you say you will do?

The Wisdom: Jeff Bezos once said, *"Your brand is what other people say about you when you're not in the room."* In the P-R-O-D-U-C-T framework, your integrity is your brand. If you don't perform, the room goes silent.

- **REVOLUTIONIZE:** Are you evolving? Or are you running "Legacy Code"? If you haven't learned a new skill in 12 months, you are becoming obsolete.

- **OPTIMIZE:** Are you easy to work with? Or do you create friction? High-value people remove complexity for others.

- **DURABLE:** Do you have resilience? Can you handle a setback without crashing?

- **USABLE:** Do you have the right Skillset and an aptitude to apply those skills in the right place?

- **CONSISTENT:** is your performance repeatable? or does it go up/down?

- **TRANSPARENT:** Do people know what they get with you? Or do you hide your motives?

The Solopreneur's Lesson: Stop worrying about your logo or your website. Focus on making *yourself* a product that screams value, reliability, and integrity.

THE AVIGNA DECISION MATRIX: Product Audit

*Before signing that contract, run the purchase through the **WWHW** filter.*

1. WHY are we buying this?

- *Blocked Answer:* "Because everyone else is moving to this platform." (FOMO).

- *Unblocked Answer:* "Because our current system cannot handle the transaction volume we project for Q4, and this specific product solves that bottleneck."

2. WHEN should we implement this?

- *The Discipline:* Do not implement a new product during your busy season. Wait until the **Prepare** phase is complete. If you rush the "When," you ruin the "What."

3. HOW will this work?

- If the vendor says, "It's plug and play," they are lying. Nothing is plug and play.

- *Ask:* Who is going to manage this on Day 2? If you don't have a name, you don't have a plan.

4. WHAT are we actually buying?

- Are we buying a solution? Or are we buying a subscription to a problem?

- Check the Exit Strategy. If we hate this product in a year, how hard is it to get our data out?

Your Thoughts and Self Reflection

(Write your key points and any personal experience related to the ideas presented in the chapter)

CHAPTER 4: SERVICE

The Invisible Asset

If Product is the "Thing" you buy, Service is the "Life" you breathe into it.

You can buy the most sophisticated server, the fastest firewall, or the most robust cloud subscription. But without the right service wrapping around it, that product is just an expensive paperweight.

In the IT world, "Service" has a branding problem. When people hear the word, they usually think of a "Help Desk"—someone you call when things break. They think of it as **Servitude**. But in the **Avigna Methodology**, Service is not servitude. It is **Partnership**.

The Service Trap Many businesses fall into what I call the "Service Trap." It is a race to the bottom driven by **Fear**.

- **The Buyer's Trap:** They look at the "Maintenance and Support" line item on a quote and see wasted money. They grind the vendor down on price, ignoring the Service Level Agreement (SLA). Then, when a crisis hits at 2:00 AM on a Saturday, they realize that "saving money" is costing them a fortune in downtime.

- **The Seller's Trap:** They view service as an afterthought—a checkbox to get the deal signed. They sell the dream (The Product) but deliver a nightmare (The Support).

Real Intelligence (RI) understands that Service is an **Asset**. It is the insurance policy for your success. It is the difference between a minor hiccup and a business-ending disaster.

When I evaluate a partner, build a team, or design a solution, I use the **S.E.R.V.I.C.E.** framework to ensure we are building a relationship that lasts.

S–SUITABLE: Fit for Purpose

Suitability is about context. A service must fit the specific needs of the customer *right now*.

I often see vendors pushing "Platinum Level" service packages to small businesses that only need basic support. Conversely, I see global enterprises trying to run 24/7 operations on a "Basic" tier to save a few pennies. Both are recipes for failure.

I worked with a client who had just invested in a major network upgrade. The technology was solid. But during the procurement phase, they decided to "opt out" of the maintenance and professional services package. The logic was simple: they wanted to stay within the annual budget. "Maintenance," they believed, could be added later "if needed."

For a while, everything appeared fine. Then one afternoon, a critical component failed. It wasn't dramatic—just a hardware fault—but it brought a key part of their business to a halt. When they reached out to the vendor, they learned the uncomfortable truth: without a maintenance contract, the replacement would take weeks, not hours.

That moment became an expensive lesson. What initially looked like a cost-saving strategy ended up costing far more—in money, time, and stress.

The Avigna Rule: Don't sell a Ferrari to someone who needs a pickup truck. And don't buy a pickup truck if you are trying to race in Formula 1. Suitability requires the discipline to say "No" if the fit isn't right.

E–EFFICIENT: Friction Reduction

Efficiency is often confused with "speed," but they are different. Speed is moving fast; Efficiency is moving without waste.

- **Inefficient Service:** You call support, wait on hold, explain your problem, get transferred to Level 2, explain it again, and get a ticket number. You have spent an hour and achieved nothing.

- **Efficient Service:** The provider anticipates the issue (via monitoring tools) and fixes it before you even call.

Efficiency respects the most valuable asset your client has: their **Energy**. If your service drains their energy, they will eventually fire you, no matter how good your product is.

R–RELIABLE: The Boredom Factor

I said this in the Product chapter, but it applies doubly here: **Reliability is boring.** And in IT, boring is beautiful. Reliability means the service is available when the sun is shining *and* when the storm is raging.

Fear creates instability. When a provider is operating out of fear (fear of losing margin, fear of staffing costs), they cut corners on reliability. They rely on "Best Effort." But "Best Effort" is not a strategy. **Guarantees** are a strategy.

V–VERIFIABLE: Trust, but Verify

Service is intangible. You cannot hold it in your hand. This is why it is so hard to sell—and so easy to fake. To counter this, good service must be **Verifiable**.

- **The Metrics:** Do you have clear SLAs (Service Level Agreements)?

- **The Data:** Can you prove the value?

If you cannot measure it, it doesn't exist. A Mindful Service provider doesn't hide the metrics; they showcase them, even when the numbers are bad. Transparency builds trust faster than perfection does.

I–INTERACTIVE: The Human Connection

This is where the **People** leg of our PPT framework shines. Service is a conversation. It is **Interactive**.

In the age of AI chatbots and automated ticketing systems, we are losing the art of interaction. We treat customers like "Ticket #9911" instead of human beings. **Real Intelligence (RI)** demands empathy. It demands that we listen more than we speak.

- **Transactional Service:** "I fixed the router. Ticket closed."

- **Interactive Service:** "I fixed the router, and I noticed your traffic spikes on Tuesdays. Let's talk about how to optimize that so it doesn't happen again."

Several years ago, I was asked to step into a project where the previous Subject Matter Expert (SME) had been removed. The client's site lead was a former contractor who knew exactly how to push your buttons. He had little patience for delays, and tension was high.

In situations like this, technical expertise only gets you so far. So instead of matching his frustration, I made a conscious decision to listen. I asked him what his real concerns were. I let him vent. I approached every task with the mindset of "How can I make his job easier?"

Over time, the tension faded. We began working as a team instead of two sides pushing against each other. That experience reminded me of a simple truth: **Sometimes the most effective technical skill on a project is listening.**

C–CAPABLE: The Competence Check

This is the baseline. Does the team actually know what they are doing? Capability is built on **Knowledge** (from our W.O.R.K. framework). It requires certification, training, and experience.

But it also requires the capability to admit when you *don't* know.

- **The Fearful Partner:** Guesses the answer to avoid looking stupid (and breaks the system).

- **The Capable Partner:** Says, *"I don't know, but I will find out,"* and comes back with the right answer.

E–EXCELLENCE: The Culture

Excellence is not an act; it is a habit. It is the culture of going one step further than the contract requires. Mediocrity is crowded. Excellence is lonely. If you commit to excellence in service, you have very little competition.

THE LIFECYCLE OF EXCELLENCE: PPDIOO

So, how do we deliver this S.E.R.V.I.C.E? We don't guess. We follow a lifecycle.

In the networking world (championed by Cisco), there is a standard service methodology known as **PPDIOO**: **P**repare, **P**lan, **D**esign, **I**mplement, **O**perate, **O**ptimize. While this is an industry standard, most people get it wrong. They treat it as a linear list (Step 1 to Step 6). In reality, it is an **Infinity Loop of Continuous Improvement**.

Here is the **Avigna Take** on why this lifecycle matters:

1. PREPARE (The Strategy Phase)

Most projects fail here. Why? Because people rush to "Design". The **Prepare** phase establishes the business case. It asks: *Why are we doing this? Who is it for? What is the budget?* If you skip preparation because you are in a hurry (Fear), you will spend 10x the time fixing mistakes later.

2. PLAN (The Assessment)

This is where we audit the current infrastructure. We assess the gap between "Where we are" and "Where we want to be."Planning manages the resources, the timelines, and the risks. It prevents the project from becoming a "money pit."

3. DESIGN (The Blueprint)

This is the architectural phase. A good design is detailed, documented, and reviewed. Crucially, this is where we align with the **PPT** framework. We don't just design the technology; we design the workflow for the **People** who will use it.

4. IMPLEMENT (The Build)

This is where the rubber meets the road. If the first three steps (Prepare, Plan, Design) were done right, Implementation should be boring. If Implementation is "exciting" (full of fires, errors, and surprises), it means you failed to Prepare.

5. OPERATE (The Day 2 Reality)

This is where the system goes live. This is "Day 2." Most vendors celebrate at "Implement" and then leave. But **Service** lives in "Operate." This is monitoring, managing, and resolving issues. This is where 90% of the lifecycle cost exists.

6. OPTIMIZE (The Kaizen Loop)

This is the most important step, and the one most frequently ignored. Once the system is running, we don't stop. We ask: *How can we make this better?*

- **The Fear-Based Provider:** "It works, don't touch it."

- **The Solution Architect:** "It works, now let's make it E.A.S.I.E.R."

Optimization feeds back into Preparation, starting the cycle anew. This is how you create long-term value.

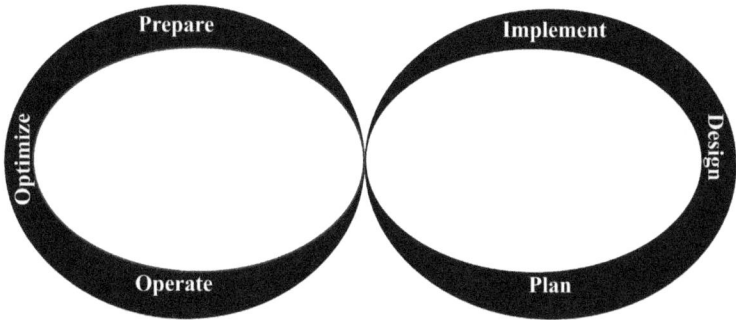

FIGURE: THE PPDIOO INFINITY LOOP

FUTURE LOOK: Service in the Age of AI

How will Artificial Intelligence change Service?

We are about to see a flood of **AI Service Agents**. I am not talking about the clumsy "Chatbots" of the past that got stuck in loops. I am talking about Agentic AI—bots that have permission to log into your systems, diagnose a network failure, write the patch code, and deploy it before a human even wakes up.

This sounds like a utopia for **Efficiency** (The 'E' in our framework). But it poses a massive risk for **Interaction** (The 'I').

The "Silence" Trap

If the AI fixes everything silently in the background, the client never calls you. You never speak. The relationship goes dark and when a relationship goes dark, you become a utility bill. You become a commodity. And commodities are always replaced by the cheapest option.

The Avigna Prediction: The "Human Premium"

As technical execution becomes automated (and free), the value of **Human Connection** will skyrocket.

In a world where everyone has access to the same AI tools, "fixing the problem" is no longer a competitive advantage. It is table stakes.

The differentiator will be **Real Intelligence (RI)**.

- **AI handles the "Break/Fix":** It resets the passwords, patches the servers, and monitors the logs. It handles the Past (what happened) and the Present (what is happening).

- **RI handles the "Strategy":** You handle the Future. You sit down with the client and ask, "Now that the network is stable, how do we use it to expand into the Asian market next year?"

The New Service Model

The goal is not to use AI to replace your service team (cutting costs). The goal is to use AI to **elevate** your service team. Instead of hiring 10 people to answer phones (reactive), you hire 10 people to be "Customer Success Architects" (proactive).

They don't spend their day saying, "Have you tried turning it off and on again?"

They spend their day saying, "I noticed your usage patterns are changing. Let's adjust your architecture to save you money."

That is not just Service; that is **Partnership**. That is how you become indispensable in the Age of AI.

THE PERSONAL PIVOT: Service as a Way of Life

In the corporate world, Service is a department. In your personal life, Service is a **Strategy**.

Many entrepreneurs fail because they view relationships as transactions. *"What can I get from this person?"*

Real Intelligence flips the script: *"How can I serve this person?"*

- **S–SUITABLE:** Are you giving people what *they* need, or what *you* want to give? (Think about your spouse, your kids, or your clients).

- **I–INTERACTIVE:** Are you truly listening? In a world of digital noise, giving someone your undivided attention is the highest form of service.

The Wisdom: Maya Angelou taught us a valuable lesson that applies to just about anyone in regards to communications: ***"I've learned that people will forget what you said, people will forget what you did, but people will never forget how you made them feel."*** Real Intelligence is knowing that the feeling is the real deliverable in your everyday interactions with your customers, partners, employees and on a personal level family and friends.

- **R–RELIABLE:** Are you the person your friends call when things fall apart? That is the ultimate metric of personal value.

The Lesson: If you want to build a network that generates unlimited opportunity, stop networking. Start serving. Be the most useful person in the room.

THE AVIGNA DECISION MATRIX: Service Audit

Before signing a support contract or hiring a service team, ask these four questions.

1. WHY do we need Premium Support?

- *Blocked Answer:* "Because the salesperson said we did."

- *Unblocked Answer:* "Because the cost of one hour of downtime exceeds the cost of the annual contract." (This is **Economical** thinking).

2. WHEN do we escalate?

- Do you have a defined process for when a problem moves from Level 1 to Level

3? Or do you wait until the customer is screaming?

- *The Rule:* Escalate based on **Impact**, not just time.

3. HOW do we measure success?

- Don't just measure "Time to Resolution." Measure "Time to Innocence" (proving it wasn't the network) and "Customer Sentiment."

- *The Metric:* Is the customer happier *after* the problem than they were before?

4. WHAT are we actually getting?

- Read the fine print. Does "24/7 Support" mean a human answers the phone, or that you can *submit a ticket* 24/7? There is a massive difference.

Your Thoughts and Self Reflection

(Write your key points and any personal experience related to the ideas presented in the chapter)

CHAPTER 5: PROFIT

The Scoreboard of Value

In polite society, people are often uncomfortable talking about money. In the business world, we obsess over it, yet we often misunderstand it.

There is a pervasive myth that **Profit equals Greed**. We hear about companies firing loyal staff to boost stock prices, or vendors gouging customers to hit quarterly targets. This is "Fear-Based Profit." It is driven by the fear of scarcity.

But in the **Avigna Methodology**, we define Profit differently.

Profit is Oxygen.

Without oxygen, you cannot breathe. Without profit, your business cannot survive. You cannot hire the best **People**. You cannot improve your **Process**. You cannot buy the best **Technology**. Profit is not the destination; it is the fuel that allows you to reach the destination.

The Profit Paradox

Here is the paradox I learned during my time studying financial markets: **If you chase money, you will lose.** If you focus solely on squeezing every dollar out of a deal, you will cut corners. You will destroy your reputation. You will lose the customer.

But if you chase **Value**—if you focus on solving the customer's problem better than anyone else—the profit becomes inevitable. **Real Intelligence (RI)** understands that Profit is a byproduct of Excellence.

To ensure your business is not just making money but building a sustainable future, I use the **P-R-O-F-I-T** framework.

P–POSITIONING: The Identity Check

Profit starts before you sell a single unit. It starts with **Positioning**.

Positioning is not just a marketing term; it is an identity. Who are you in the mind of the customer?

There are generally two places to be:

1. **The Low-Cost Leader:** You sell commodities. You win on price. (Think Walmart).

2. **The Premium Value:** You sell solutions. You win on results. (Think Apple or a specialized neurosurgeon).

The "Danger Zone" is the middle. The middle is where you are too expensive to be cheap, but not good enough to be premium. **Fear** drives companies to the middle. They are afraid to charge what they are worth, so they discount. But they are afraid to cut costs, so they bloat.

One thing I've noticed—both in my own line of business and while working with clients in different industries—is that many organizations struggle to define who they really are. They try to be everything to everyone, hoping that a wider net will bring more opportunities. Instead, they end up diluting their value.

I once worked with a small technology firm that started out with a very clear niche. They were experts in a particular network technology. But over time, they began adding more services such as development, staffing, even graphic design. The catalog grew, but their clarity shrank.

Customers no longer knew what the firm stood for. Eventually, they became seen as a "body shop," supplying people rather than specialized expertise. Their margins dropped, and their reputation became generic. The market defined them because they failed to define themselves.

The Avigna Rule: Focus does not limit you—focus strengthens you. Pick a lane. If you offer Premium Value, do not apologize for your price. Profit comes from clarity.

R–REVOLUTIONARY REVENUE:
Recurring Value

Notice I did not just say "Revenue." I call it **Revolutionary Revenue**.

- **Traditional Revenue** is transactional: *I sell you a box, you give me cash. Goodbye.*

- **Revolutionary Revenue** is relational: *I solve your problem today, tomorrow, and next year.*

In the IT world, this is the shift from "Break/Fix" (waiting for things to break) to "Managed Services" (preventing things from breaking). It is the shift from buying software once to "SaaS" (Software as a Service).

Why does this matter for profit? Because **Customer Retention** is cheaper than **Customer Acquisition**. If you have to hunt for a new customer every month, your profit margins will be eaten by marketing costs. If your revenue is recurring, your profit compounds.

O–OPERATIONAL OVERHEAD: The Lean Philosophy

This is the "Cost" side of the equation. In accounting terms, Operational Overhead is the money you spend just to keep the lights on (Rent, Admin, Utilities). In strategic terms, Overhead is the enemy of agility.

However, be careful. There is a difference between **Cutting Fat** and **Cutting Muscle**.

- **Cutting Fat:** Eliminating waste, automating manual tasks, renegotiating software licenses. This is healthy.

- **Cutting Muscle:** Firing your best engineers or stopping innovation to hit a quarterly number. This is fatal.

I once worked with a mid-sized organization that decided to slash its training budget—not trim it but eliminate it entirely. The thinking was simple: *"There is so much free training online. Our engineers can self-learn."* On paper, it looked like a cost-saving win.

Then came the outage. A routine change turned into a major incident when an engineer—who had pieced together his "training" from random blog posts—executed a configuration that wasn't compatible with the new architecture.

The result? A network outage that lasted hours. The cost of that single outage exceeded the entire year of training budget they had cut. They realized too late that **training is not an expense; it is insurance.**

F–FLEXIBILITY: The Survival Skill

The market is a living organism. It changes.

- In 2019, remote work was a luxury. In 2020, it became a necessity.

- In 2022, AI was a novelty. Today, it is a requirement.

If your business model is rigid, you will break. **Flexibility** protects your profit. This means having the ability to pivot your offering without tearing down your infrastructure. It connects back to the

Agile component of our Technology framework. A flexible company can weather a storm that sinks a rigid one.

I–INNOVATION: The Growth Engine

If you are not moving forward, you are moving backward. Inflation alone ensures that standing still is a loss.

Innovation is how you protect your margins.

When a product is new, margins are high. As time goes on, competitors copy you, and it becomes a commodity. Margins drop. To keep your profit healthy, you must constantly innovate.

- *Note:* Innovation doesn't always mean inventing the iPhone. It often means **Continuous Improvement** (Kaizen). Can we deliver this service 10% faster? Can we reduce errors by 5%? Small innovations compound into massive profits.

T–TRANSFORMATION: The Ultimate Goal

Finally, what do you do with the profit?

You **Transform**.

This is the cycle of business life.

1. **Input:** You invest capital and effort.

2. **Process:** You apply People and Technology (PPT).

3. **Output:** You create Value.

4. **Profit:** You capture value.

5. **Transformation:** You reinvest that profit to build a better business.

If you take the profit out of the business to buy fancy yachts and ignore the business, the cycle stops. That is greed. If you use the profit to train your people, upgrade your tools, and refine your process, the cycle accelerates. That is **Sustainability**.

CONSULTANT'S INSIGHT: Financial Literacy for Tech Leaders

One of the biggest obstacles I see is that many IT leaders and operational managers do not understand basic financial literacy. They think "Profit" is the CFO's job.

It is your job.

Every decision you make—hiring a person, buying a server, changing a process—is a financial decision.

The "Efficiency Ratio"

You don't need to be a CPA, but you must understand one ratio:

- *How much does it cost us to generate $1.00 of revenue?*

If you are spending $0.90 to make $1.00, you are fragile. One bad month will wipe you out. If you are spending $0.60 to make $1.00, you are robust. You have a war chest. You can survive a crisis.

I do want to emphasize here that this framework is a strategic guide, not a substitute for comprehensive financial planning. I do recommend that leaders use this framework to guide their *thinking* about profit but should always work with qualified financial professionals for detailed analysis. Do not ignore the math. Use the **P-R-O-F-I-T** framework to drive that cost down (through Efficiency) and drive that revenue up (through Positioning), and you will build a business that is not just profitable, but unstoppable.

FUTURE LOOK: Profit in the Age of AI

How will Artificial Intelligence change the bottom line?

For the last decade, managing IT costs—especially Cloud costs—has been a nightmare. We moved from buying servers once every three years (Capital Expense) to renting them by the second (Operating Expense). The result? Complexity exploded. Companies are wasting millions on "Zombie Infrastructure"—servers that were spun up for a test, forgotten, and left running on the company credit card for months.

In the next 5 years, we will see the rise of **Autonomous FinOps** (Financial Operations). We won't just have dashboards telling us what we spent;

we will have AI Agents that actively manage the wallet.

The Opportunity: Surgical Precision

AI is the perfect tool for cutting "Fat. " An AI agent can monitor 10,000 cloud instances simultaneously. It can spot a server that is idling at 2:00 AM and shut it down, then spin it back up at 7:55 AM right before the team logs on. It can predict that you will need more storage next month and pre-purchase it at a discount. This level of real-time optimization is impossible for a human. It will save companies billions.

The Risk: Context-Blind Cutting

However, there is a massive trap. **AI optimizes for Efficiency, not Effectiveness.** AI looks at a spreadsheet, sees a cost with no immediate revenue attached, and flags it as "Waste."

- **The Scenario:** You have a training budget. Or an R&D sandbox where engineers are experimenting with a new idea that hasn't launched yet.

- **The AI View:** "This department is spending $10,000/month and generating $0.00 revenue. Recommendation: Cut budget."

- **The Reality:** That R&D sandbox is your future product. That training budget is what keeps your staff from quitting.

If you let AI run your profit strategy on auto-pilot, it will cut the "Muscle" because it looks like "Fat" on a spreadsheet. It will optimize you into irrelevance.

The Role of Real Intelligence: The Value Audit

Real Intelligence (RI) is required to provide the Context. Your job as a leader is to teach the AI what "Value" means.

- You tell the AI: "Do not touch the Innovation Lab budget. That is a strategic investment."

- You tell the AI: "Optimize the legacy server, but leave the customer support portal alone because speed is more important than cost there."

The Prediction: Profitability will no longer be about who can cut costs the fastest. Everyone will have AI to do that. Profitability will be about who has the wisdom to know **where to spend**. The companies that win will be the ones that use AI to save money on the boring stuff, so they can pour that money into the **People** and **Innovation** that actually drive growth.

THE PERSONAL PIVOT:
Energy Economics

In business, we measure Profit in dollars. In life, we measure Profit in **Energy** and **Freedom**.

Every day, you wake up with a bank account of Energy.

- **Revenue:** Activities that recharge you (Creativity, Connection, Wins).

- **Overhead:** Activities that drain you (Commuting, Arguing, Worrying).

- **Profit:** The energy you have left over to build your future.

The Bankruptcy Trap: Many high-performers are running at a loss. They have high financial revenue, but their "Emotional Overhead" is so high they are mentally bankrupt. They are burning muscle to stay warm.

The Wisdom: The stoic philosopher Seneca wrote, *"It is not that we have a short time to live, but that we waste a lot of it."* When you waste your energy on high-overhead emotions (worry, anger, ego), you are literally spending your life currency on things that give zero return.

The Fix:

Apply the **P-R-O-F-I-T** model to your calendar.

- **Cut the Overhead:** Say "No" to obligations that don't align with your values.

- **Invest in Transformation:** Take your "Profit" (your free time) and reinvest it in learning, health, and relationships. Don't just spend your time; invest it.

THE AVIGNA DECISION MATRIX: Profit Audit

*Run your business through the **WWHW** filter to see if your profit is healthy or toxic.*

1. WHY are we profitable?

- *Blocked Answer:* "Because we cut costs to the bone." (This is unsustainable).

- *Unblocked Answer:* "Because customers value our solution and stay with us for years." (This is value-driven).

2. WHEN do we reinvest?

- *The Trap:* Hoarding cash when the market is shifting.

- *The Discipline:* Reinvesting when you are *ahead*, not scrambling to invest when you are falling behind.

3. HOW do we measure success?

- Do you look at "Gross Revenue" (Ego metric)?

- Or do you look at "Net Profit" and "Customer Lifetime Value" (Reality metrics)?

4. WHAT is our differentiator?

- If you can't name the one thing that justifies your price, you are a commodity. Go back to **Positioning**.

Your Thoughts and Self Reflection

(Write your key points and any personal experience related to the ideas presented in the chapter)

PART 3

Let's Modernize

CHAPTER 6: WORK

Redefining the Grind

I want to start this chapter with an interaction I had with a store manager of McDonalds in 1990, the year I first came to America as a twenty-year-old student. I needed a part-time job to earn some pocket money and through a reference I ended up visiting a nearby McDonalds to meet the store manager. He knew that I was a newbie in need of a side gig. He explained me the duties that included tasks like dishwashing, floor cleaning and mopping along with customer-facing duties that would come as I gained more experience. During his serious pitch of the job, as any good manager would, I am sure he noticed the awkward look on my face and said "**No job is too small**" here in America. You can make money as long you do whatever needs to be done to take care of your customers, he added. I was too naïve to understand what he really meant back then and I did not take the job as I did not come to America to wash dishes. But that sentence **"No job is too small"** stuck with me somehow and as I matured, I embraced it as my personal philosophy in the work I currently do.

Moving on, we live in a culture that creates a false dichotomy about work.

On one side, we have the **"Grind"** culture. Social media influencers tell you to wake up at 4:00 AM, take ice baths, and work until you collapse. They sell the idea that suffering is a badge of honor. On the other side, we have the **"Hack"** culture. These are the people who claim you can outsource everything to AI, work four hours a week, and become a millionaire without lifting a finger.

Both views are flawed because they view work as a *cost*.

In the **Avigna Methodology**, we view work as an **investment**.

When I look back at my career—from network engineering to financial risk management—the work wasn't just about the paycheck. It was about the **Education**. Every project, every crisis, and every late night was a deposit into my bank of wisdom.

To succeed, we need to redefine what "Work" actually is and change our relationship with "Hard Work."

What is W.O.R.K?

Most people think work is "Trading time for money." I define it differently: **W.O.R.K is Willing to Obtain Real Knowledge.**

When you view work through this lens, you stop being a laborer and start being a student. The outcome of the work isn't just the product you deliver to the client; it is the knowledge you keep for yourself.

W – WILLING

This is the starting point. You must be **Willing**. Willingness is not just "compliance." It is an active state of curiosity. It is raising your hand for the difficult assignment because you know it will teach you something.

- **The Fear-Based Employee:** "I hope they don't pick me for that project; it looks hard."

- **The Mindful Employee:** "I am willing to take this on because I want to learn how to solve this problem."

O–OBTAIN

Knowledge isn't given; it is **Obtained**. You have to go get it. In the IT world, you can read the manual (Theory), but you don't truly "Obtain" the knowledge until you break the server and fix it. You obtain wisdom through the friction of doing.

When I started my journey 32 years ago, I didn't have a fancy job title or a corporate budget. I just had curiosity. I wanted to understand networking, but reading about it wasn't enough.

So, I turned my home into a laboratory. I scraped together the hardware to build a two-computer network just so I could install **Novell NetWare**. I spent nights troubleshooting connection issues, figuring out IRQ conflicts, and staring at command

lines until my eyes blurred. I didn't stop until
I earned my first certification: **Certified Novell
Administrator (CNA)**.

That humble two-computer setup was the seed
for everything that followed. It taught me that
Knowledge isn't handed to you in a classroom; it is
Obtained by getting your hands dirty.

R—REAL

We are not interested in theoretical knowledge.
We want **Real** knowledge. Academic knowledge is
knowing *how* a firewall works. **Real** knowledge is
knowing what to do when the firewall fails at 2:00
AM. Work is the only way to convert "Book Smarts"
into "Street Smarts."

K—KNOWLEDGE

This is the ultimate asset. The money you earn can
be spent. The title you hold can be taken away. But
the **Knowledge** you obtain through work is yours
forever. Also known as Experience.

The Truth About H.A.R.D Work

Now, let's address the elephant in the room: **Hard Work.**

The word "Hard" has a negative connotation. It sounds like suffering. It sounds like a struggle. But in my philosophy, **H.A.R.D** is not a measure of difficulty; it is a measure of **Character**.

I use the acronym **H.A.R.D** to describe the state of mind required for excellence:

Happy–**A**ttitude–**R**igor–**D**iscipline.

H–HAPPY (The Joy of the Craft)

This surprises many people. How can hard work be happy? Think about a master craftsman carving wood, or a coder in the "flow state." They are working hard, but they are not suffering. They are **Happy**. Happiness here doesn't mean "giddy." It means **fulfillment**. It means finding joy in the act of solving the puzzle.

A–ATTITUDE (The Lens)

Your attitude determines your altitude. In consulting, things go wrong. Projects get delayed. Clients get angry.

- **Negative Attitude:** "Why is this happening to me?" (Victim Mindset).

- **Positive Attitude:** "How can I fix this?" (Solver Mindset).A "Hard Worker" in the Avigna sense maintains a constructive attitude even when the pressure is on.

R–RIGOR (The Standard)

Happiness and Attitude are the "Soft" skills. **Rigor** is the "Hard" skill. Rigor is the commitment to precision. It is the refusal to cut corners. It is checking the configuration one last time.

I once worked with an enterprise security team rolling out a major firewall policy overhaul. It was a high-visibility project with a tight timeline—exactly the environment where mistakes happen.

One afternoon during a peer review, I was going through the rule set line-by-line. Everything looked fine on paper. But something felt off. A specific policy was supposed to restrict access between two sensitive segments. It was configured correctly in the syntax, but my simulation showed traffic was still permitted.

I dug deeper and realized: The rule had been configured... but never applied to the active policy package. It was a "hole" that would have carried into production. When I flagged it, the room went silent. No one was careless—they were just trusting the visual. **Rigor** caught what the tools missed. It saved us from a headline-level mistake.

The Avigna Insight: Rigor isn't bureaucracy—it's protection. The cost of rigor is a few extra minutes. The cost of missing rigor can be a security breach.

D–DISCIPLINE (The Consistency)

Motivation gets you started; **Discipline** keeps you going. Discipline is the ability to maintain your Happiness, Attitude, and Rigor even when you are tired. It is showing up with the same level of excellence on Friday afternoon as you did on Monday morning.

The Multiplier (S.M.A.R.T)

If **W.O.R.K** is the fuel, and **H.A.R.D** is the engine, then **S.M.A.R.T** is the turbocharger.

In mathematics, Hard Work adds value (1+1=2). But Smart Work multiplies it (1x10=10).

We all know people who work 80 hours a week but stay in the same position for ten years. They are working Hard, but not Smart. They are digging a hole with a spoon because they are too busy to go get a shovel.

To unlock exponential growth, you must apply the **S.M.A.R.T** method to your effort.

S–STUDY: The Anti-Obsolescence Strategy

The world is changing too fast for you to rely on what you learned in college—or even what you learned last year. In the IT industry, the "Half-Life" of a skill is about 12-18 months and it is shrinking as technological innovations progress. As a case in point, the major certifications (CCIE, CISSP, CCSP, PMP etc.) that I currently hold have a three-year re-certification cycle and we are required to keep updating progress every year to keep up to date in network and security technologies. If you stop studying to upskill yourself today, you are at a risk of becoming obsolete by next year to year and a half.

Study is not just passive reading. It is active curiosity.

- **The Fear Mindset:** "I don't have time to learn AI; I'm too busy doing my job." (This is the path to irrelevance).

- **The Growth Mindset:** "If I don't study this new tool, I won't *have* a job in two years."

The Avigna Rule: Dedicate 5% of your work week to studying the future of your industry. If you work

40 hours, that is just 2 hours. That investment compounds like interest in a bank account.

M–MASTER: Moving from Competence to Intuition

Don't just dabble. Master it. There is a difference between "knowing how to use a tool" and "Mastery."

- **Competence:** You can do the task if you have the manual open.

- **Mastery:** You can do the task intuitively. You see, the solution before you even touch the keyboard.

Mastery is what allows you to work *fast*. When you master the basics (whether it's Python code, financial derivatives, or public speaking), you free up your brain's CPU to focus on **Strategy**.

Bruce Lee said it best: ***"I fear not the man who has practiced 10,000 kicks once, but I fear the man who has practiced one kick 10,000 times."***

A–APPLY: The Death of Theory

Knowledge without application is just trivia. It is "Shelf-ware" for your brain. I see many professionals who are "Workshop Junkies." They go to every conference, buy every book, and take every certification. But they change *nothing* about how they work on Monday morning.

Real Intelligence is biased toward action.

- Did you read about a new AI tool? Download it and use it on a small project *today*.

- Did you learn about a new sales technique? Try it on your next call.

The Rule: Learn on Monday, Apply on Tuesday. If you wait until Wednesday, you will forget it.

R–REVISE: The Feedback Loop

This is the "Agile" component of personal work. You will not get it right the first time. The "Fear-Based" worker is terrified of making a mistake, so they plan forever. The "Smart" worker executes, fails, and **Revises**.

Smart Work requires the humility to say, *"This isn't working. Let's change it."* Hard workers keep digging the hole even when they hit rock. Smart workers realize they are digging in the wrong spot, pack up their shovel, and move.

T–TRACK: The Scoreboard

Finally, you must **Track** your results. This brings us back to the **Verifiable** component of Service.

- *Are you just "busy"?* (Answering emails, attending meetings).

- *Or are you "productive"?* (Closing deals, shipping code, solving problems).

If you don't track it, you are just guessing. You might feel tired at the end of the day, but that doesn't mean you accomplished anything. Use data to determine where to apply your hard work.

FUTURE LOOK: Work in the Age of AI

Will AI make hard work obsolete?

This is the question keeping millions of professionals awake at night. If an AI agent can write code, draft legal briefs, and configure firewalls faster than a human, what is left for us to do?

My answer is counter-intuitive: **AI will not kill Hard Work. It will elevate it.**

The Death of Drudgery

For the last 50 years, "Knowledge Work" has actually involved a lot of drudgery.

- The Network Engineer spends 80% of their time parsing logs and 20% designing architecture.

- The Financial Analyst spends 90% of their time cleaning data in Excel and 10% making investment decisions.

AI is coming for the 80% and the 90%. It will ruthlessly automate the data entry, the syntax checking, and the routine coding. This is not a loss; it is a liberation. It frees humans to do the **"Happy"** work (from our **H.A.R.D** framework): The Strategy, the Architecture, and the Relationship building.

The New Risk: The "Auto-Pilot" Atrophy

However, there is a danger here. If we let AI do everything, we risk losing the ability to judge quality. Think of a pilot who uses auto-pilot so much they forget how to fly manually. When a storm hits and the sensors fail, that pilot is helpless.

In the Age of AI, **Rigor** becomes the most valuable skill on earth. Because AI is confident, but it is not always right. It hallucinates. It makes logical leaps that ignore context.

- **The Old Work:** Writing the code.

- **The New Work:** Having the deep expertise to look at the AI-generated code and say, "That looks correct, but it introduces a security vulnerability in the payment gateway."

The "Human Premium"

As technical execution becomes a commodity (cheap and abundant), **Human Connection** becomes a premium asset (rare and expensive). You cannot automate trust. You cannot automate the handshake that closes a deal, or the empathy required to calm an angry client during an outage.

The Prediction: The workforce will split into two groups:

1. **The Operators:** People who just feed prompts to the AI. They will be paid like machine operators.

2. **The Architects:** People who use **Real Intelligence** to question, guide, and audit

the AI. They will be paid like masters.

To stay in the second group, you must never stop **Obtaining Knowledge**. You must understand the "First Principles" of your industry so deeply that you can grade the AI's homework.

THE PERSONAL PIVOT:
The Student of Life

The concepts of **W.O.R.K** and **H.A.R.D** are the antidote to the "Imposter Syndrome" that plagues so many solopreneurs and entrepreneurs.

- **W.O.R.K:** If you view your career as a way to "make money," you will always feel anxious. If you view your career as a way to "Obtain Knowledge," you become fearless. Even a failure is a win because you obtained knowledge.

The Wisdom: Mahatma Gandhi said, *"Live as if you were to die tomorrow. Learn as if you were to live forever."* This is the essence of the Avigna definition of Work. We do not work just to earn; we work to learn.

- **H.A.R.D:** The happiest people I know are not the ones sipping cocktails on a beach. They are the ones engaged in difficult, meaningful pursuits. For them "No job is too small".

The Lesson: Don't wish for it to be just easy. Wish for the **Discipline** to enjoy the climb. That is where happiness lives.

THE AVIGNA DECISION MATRIX: Work Audit

Are you working Hard, or are you just busy?

1. WHY am I doing this task?

- *Blocked:* "Because I was told to."

- *Unblocked:* "To obtain knowledge that solves a client problem."

2. WHEN do I apply Rigor?

- *The Trap:* Applying rigor only when the boss is watching.

- *The Discipline:* Applying rigor when *no one* is watching.

3. HOW do I measure my output?

- Do not measure hours worked. Measure problems solved.

4. WHAT knowledge did I obtain today?

- If the answer is "Nothing," you wasted the day.

Your Thoughts and Self Reflection

(Write your key points and any personal experience related to the ideas presented in the chapter)

CHAPTER 7: MODERNIZATION

Making IT E.A.S.I.E.R

We have arrived at the "Technology" conversation.

In the past, technology was a department in the basement. Today, it is the nervous system of your business. If the network goes down, the business stops. If the data is breached, the reputation is destroyed.

But this importance has bred a new kind of anxiety: **The Complexity Trap.** Leaders are bombarded with buzzwords: *Multi-Cloud, Edge Computing, Kubernetes, Generative AI.* Driven by **Fear of Missing Out (FOMO)**, companies rush to buy the latest "Shiny Object."

The result? They don't get "Modernization." They get "Modern Chaos."

The Avigna Definition of CLOUD

At Avigna, we define "CLOUD" differently: **C**onsumers **L**everage **O**ver **U**sage **D**emand.

This means *you* (the Consumer) should have total leverage over the technology. You should control the demand, the cost, and the outcome. If the technology controls you, you have failed.

To regain control, we use the **E.A.S.I.E.R** method.

E—ELASTIC: The Breathing Room

Modern tech must be Elastic. It needs to breathe in, breathe out. Elasticity means the system scales up when you grow and—more importantly—scales down when you rest.

A few years back, I worked with a client transitioning to the cloud. Culturally, they were still carrying old data-center instincts. They were used to buying hardware years in advance, so they were afraid of running out of capacity.

During planning, the Director said, "Let's make everything two sizes larger. Just to avoid surprises." They over-provisioned across the board. A month after go-live, performance was excellent, but systems were idling at 10% utilization. It was like paying for a fleet of trucks when they only needed two vans. We had to sit down and "Right-Size" the instances. Once we enabled auto-scaling, the savings were immediate.

The Lesson: The shift to cloud isn't just technical—it's behavioral. Over-provisioning feels safe, but it defeats the purpose of elasticity.

A—AGILE: Velocity over Methodology

"Agile" has become a buzzword ruined by consultants. It doesn't just mean "doing daily stand-up meetings."

Agile means Velocity. It is the ability to pivot without breaking. If you cannot deploy a new feature in weeks (or days), you are already behind.

S—SECURE: The Currency of Trust

Security is not just about firewalls; it is about **Reputation Management**. Whenever I talk about cybersecurity, I remind clients that the biggest risks usually aren't hidden in code—they're sitting at someone's keyboard.

A few years back, I advised an enterprise with a solid security stack on paper. Next-gen firewalls, Zero Trust roadmap—it looked perfect. But during a phishing assessment, several users (including technical staff) clicked on fake emails. One click opened a tunnel the security tools never saw. The VP of Security looked at me and said, "We keep thinking we are one product away from safety." The real fix wasn't another software license. It was training and behavioral change. **Technology can strengthen your defenses. But the moment a human makes a bad decision, even the best tools become spectators.**

I—INTELLIGENT: The AI + RI Equation

This is the most critical shift of our generation. **Artificial Intelligence (AI) is incomplete without Real Intelligence (RI).**

AI is a prediction engine (Speed, Scale, Pattern Recognition). **Real Intelligence (You)** provides Context, Empathy, and Ethics.

The Avigna Equation:

AI (Tech) + RI (People) = Success

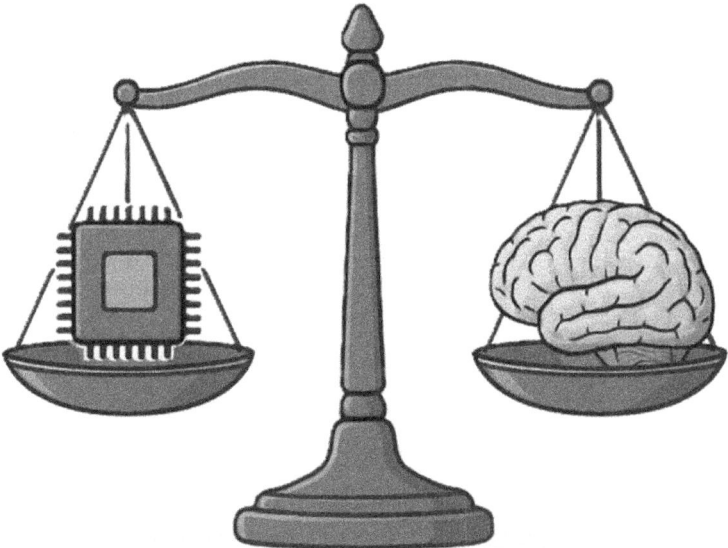

FIGURE: THE INTELLIGENCE EQUATION

Technology (AI) provides the horsepower, but People (RI) provide the steering. You need both to reach the destination.

If you deploy AI without RI, you get "Artificial Stupidity." If you rely only on RI without AI, you get "Burnout."

Modernization Strategy: Use AI to handle the data, so your Real Intelligence can handle the decisions.

E—ECONOMICAL: Value over Cost

Treat technology spend like an investment portfolio (**Financial Operations**). Every dollar spent on IT must have a direct line to business value.

R—RISK REDUCER: Resilience

Finally, technology must reduce risk, not create it. A resilient architecture builds redundancy. It assumes failure will happen and designs a way to survive it.

FUTURE LOOK: The Autonomous Enterprise

Where is this going?

We are moving rapidly toward the era of **Self-Healing Networks** and **AIOps** (Artificial Intelligence for IT Operations).

In the old world (the Novell/Cisco days), if a link went down, a pager went off. An engineer woke up, logged in via CLI, identified the break, and typed a command to reroute traffic.In the **Autonomous Enterprise**, the AI detects the latency spike before the link even fails. It automatically spins up a new instance in a different region, reroutes the traffic, and patches the vulnerability—all while the engineer is asleep.

This sounds like a utopia where humans are obsolete. **It is the opposite.**

The "Flash Crash" Risk In financial markets, we saw the rise of High-Frequency Trading (HFT). Algorithms bought and sold stocks in milliseconds. It created massive efficiency.But it also created the "Flash Crash." If one algorithm made a mistake, other algorithms reacted to it, creating a cascade of failure that wiped out billions of dollars in minutes.

The same risk exists in the Autonomous Enterprise.If an AI decides that a specific traffic pattern looks like a threat, it might shut down your entire e-commerce portal to "save" it. It executed the task perfectly, but it killed the business logic.

The Role of Real Intelligence: "Intent-Based Networking"This makes **RI (Real Intelligence)** more critical than ever. We are moving from "Configuring Devices" to "Defining Intent."

- **The Old Job:** "Type this code into the router."

- **The New Job:** "Teach the AI that

Customer Experience is more important than *Bandwidth Cost* during Black Friday."

We need humans to design the **Logic**, the **Ethics**, and the **Guardrails** that the AI follows.An autonomous car can drive itself, but a human must tell it where to go. If you don't input the destination, the car is just a very expensive piece of metal sitting in the driveway.

The Prediction: The IT engineers of the future will not be "technicians." They will be **Governance Architects**. They will spend their days auditing the AI's decisions to ensure they align with the business goals (PROFIT) and human values (PEOPLE).

THE PERSONAL PIVOT: Your AI Strategy

You are hearing a lot about AI replacing jobs.

- *The Fear:* "A robot will take my job."

- *The Reality:* "A human using **Real Intelligence** + AI will take your job."

You must apply the **E.A.S.I.E.R** methodology to your own skill set.

- **Agile:** Can you unlearn old skills and relearn new ones quickly?

- **Intelligent:** Are you using AI to handle your "commodity tasks" (emails, scheduling, basic research) so you can focus on your "Expertise"?

The Solopreneur's Edge

If you are a team of one, AI is not a threat; it is your staff. It is the infinite intern. But it needs a Manager. **You are the Manager.** Your job is no longer to *do* the thing; your job is to have the Taste, Judgment, and Ethics to direct the AI to do the thing. That is **Real Intelligence**.

The Wisdom: Charles Darwin's lesson is more relevant now than ever:

"It is not the strongest of the species that survives, nor the most intelligent that survives. It is the one that is most adaptable to change."

AI is the change. Be the species that adapts.

THE AVIGNA DECISION MATRIX: Tech Audit

Run your next tech purchase through the E.A.S.I.E.R filter.

1. WHY are we modernizing?

- *Blocked:* "Because everyone else is."

- *Unblocked:* "To gain Elasticity and Agility."

2. WHEN do we use AI?

- *Rule:* Use AI for data. Use Humans for decisions.

3. HOW do we secure it?

- Assume the breach has already happened. How fast can you recover?

4. WHAT is the exit strategy?

- If this vendor doubles their price next year, can we leave? If not, you are not Elastic.

Your Thoughts and Self Reflection

(Write your key points and any personal experience related to the ideas presented in the chapter)

CHAPTER 8: FOCUS

Finding Opportunities Creating Unlimited Success

We have reached the final pillar.

Let's look at what we have built so far.

- We have established the **Foundation** (PPT).

- We have defined the **Metrics** (P-R-O-D-U-C-T, S-E-R-V-I-C-E, P-R-O-F-I-T).

- We have built the **Engine** (W.O.R.K & H.A.R.D).

- We have modernized the **Tooling** (E.A.S.I.E.R).

But a car, no matter how powerful the engine, is useless without a steering wheel. That is where **F.O.C.U.S** comes in.

In my years of consulting, I have seen brilliant leaders fail because they were "busy." They confused activity with achievement. True success requires the discipline to narrow your vision so you can expand your results.

I define F.O.C.U.S as an active strategy: **F**inding **O**pportunities **C**reating **U**nlimited **S**uccess.

F—FINDING: The Hunter's Mindset

Opportunity doesn't knock; it hides. You must have the hunter's mindset to find the signal in the noise. Don't scan for threats (Fear); scan for unmet needs.

O—OPPORTUNITIES: The Reframe

A "fear-based" person sees a crisis; a "focused" person sees a chance to solve a new problem. **Real Intelligence** looks at a bug and sees the opportunity for a partnership. Opportunity is in the context.

C—CREATING: The Antidote to Fear

Finding an opportunity is potential. **Creating** is kinetic.

Early in my consulting career, I learned something important: fear grows fastest when you're standing still. I was working on an early deployment of a new networking technology. The product was so new that the documentation was thin. The client was nervous. The standard playbooks didn't work. So I did the only thing that works: I started creating one. I built lab simulations. I drafted a deployment blueprint from scratch. Once I put the first version in front of the client, everyone relaxed. **Fear thrives in ambiguity. It disappears the moment you start building.**

U—UNLIMITED: The Mindset

Why do I choose the word "Unlimited"? This is personal to me.

My first name, **Asim**, translates to **"Infinite"** or **"Unlimited."** Growing up, I viewed this not just as a name, but as a standard to live up to.

In business, we are taught to look for constraints. We are taught that the budget is limited, the market is limited, and our time is limited. But **Real Intelligence (RI)** understands that limitations are almost always self-imposed.

- **Technology Limits:** Cloud Elasticity makes our infrastructure unlimited.

- **Process Limits:** Agile makes our velocity unlimited.

- **Mindset Limits:** This is the final barrier.

If you believe profit is just "cutting overhead," your success is limited. But if you believe profit is "Innovation," your upside is **Unlimited**.

S—SUCCESS: Alignment

Finally, what is Success? In this book, we have redefined it. Success is not just the deal closed or the commission check cashed.

Success is Alignment.

- It is the alignment of **People, Process, and Technology**.

- It is the alignment of **Artificial Intelligence** and **Real Intelligence**.

- It is the alignment of **Buyer** and **Seller**—moving from a transaction based on fear to a partnership based on trust.

FUTURE LOOK: The Focused Leader

In the future, "Attention" will be the scarcest commodity on earth. AI can generate infinite content. It can generate infinite noise. The leader who can **F.O.C.U.S**—who can cut through that noise and find the signal—will be the most valuable person in the room.

In 1971, the Nobel Prize-winning economist Herbert Simon famously said, ***"A wealth of information creates a poverty of attention."***

He said that before the Internet. Before the iPhone. Before ChatGPT.

Today, we are entering the era of Infinite Content.Generative AI has reduced the cost of creating content to zero.

- An AI can write 1,000 emails in a minute.

- It can generate 50 strategic reports in an hour.

- It can create infinite charts, graphs, and slide decks.

The Consequence: If you are a leader who operates by "gathering information," you are doomed. You will drown. You cannot possibly read, process, or verify the flood of data coming your way. The "Hard Working" leader tries to read everything and stays late responding to emails. They burn out. The "Focused Leader" realizes that their job description has changed.

From "**Creator**" to "**Curator**"

In the next decade, the most valuable skill in the C-Suite won't be generation; it will be Filtration.

- Artificial Intelligence creates the noise.

- Real Intelligence (RI) provides the filter.

The Focused Leader is the one who can walk into a boardroom filled with AI-generated projections and ask the one question that matters. They are the one who can turn off the notifications, close the laptop, and think deeply about a single problem for two hours.

Deep Work as a Competitive Advantage because attention is scarce, it is expensive. If you can maintain F.O.C.U.S.—if you can hold a single strategic thought without being distracted by the "shiny object" of new data—you possess a superpower. While your competitors are reacting to every trend the AI flags, you are executing on the one truth that remains constant.

The Prediction: This could be controversial, but we will soon see "low-tech" sanctuaries in high-performance companies. We will see "No-Device Meetings" becoming the standard for high-stakes decisions. The leaders who command the highest fees will not be the ones who are "always on"; they will be the ones who have the discipline to turn it off.

THE PERSONAL PIVOT: The Unlimited Self

We started this book talking about "Unblocking." The final block is almost always **Self-Imposed**.

We tell ourselves stories about our limits:

- *"I'm too old to start a business."*

- *"I'm not technical enough to understand AI."*

F.O.C.U.S. is the tool to break those stories.

- **Finding Opportunities:** Stop looking at your limitations. Start looking at the market's needs.

- **Creating:** Action cures fear. When you

create value for one person, you prove to yourself that you are capable.

- **Unlimited:** Your potential is only limited by your willingness to learn (W.O.R.K.) and your ability to align your Real Intelligence with the tools of today.

The Wisdom: Henry Ford famously said, *"Whether you think you can, or you think you can't, you're right."* The only difference between a "Blocked" career and an "Unlimited" one is the story you tell yourself about what is possible.

You are the Solution Architecture of your own life. Go build something Unlimited.

THE AVIGNA DECISION MATRIX: F.O.C.U.S Audit

1. WHY are we doing this project?

- If the answer is "Because we've always done it," kill the project.

2. WHEN do we say No?

- Focus is not about what you say Yes to; it is about what you say No to.

3. HOW do we measure alignment?

- Are Sales, IT, and Ops telling the same story?

4. WHAT is the Unlimited Vision?

- If you weren't afraid of failing, what would

you build today?

Your Thoughts and Self Reflection

(Write your key points and any personal experience related to the ideas presented in the chapter)

CONCLUSION

The Avigna Promise

We have covered a lot of ground in these pages. But if you take only one thing away from this book, let it be this: **You are the Solution Architect of your life.** No matter where you are currently in your journey, you can always start from a place of thinking constructively viewing a problem as an opportunity to find solutions.

It was my sincere attempt to show my readers how you can think differently by using an acronym-based chapter structure that caters to both business leaders and individuals. I live in a world of acronyms, and this idea was conceived a long time ago when I was dealing with my own insecurities in tech and wanted to explore greener pastures in the world of financial markets. The unplanned path of studying human psychology hit the right chords in my blocked mind and gave me the opportunity to look deeper within and discover my own blockages. I am grateful to have found ways to realign both my left and right brain during this self-discovery, and, as one of the positive outcomes, I was able to find the courage to write this book for you.

Throughout my career, I have learned that the biggest obstacles are rarely technical. Technology works. Servers run. The code compiles. The obstacles are almost always us humans as we are the operators. Our behavior in particular is governed by our thoughts and belief systems that we acquire as part of our upbringing and social environments. Some important ones we discussed:

- **Fear** is an obstacle.

- **Ego** is an obstacle.

- **Complexity** is an obstacle.

You now possess the tools to remove them.

- You have the **PPT** balance. You can now think in 3 dimensions.

- You have the **E.A.S.I.E.R** roadmap. You can now walk your path to success.

- You have the **F.O.C.U.S** to direct your energy for your growth.

The framework I have provided is just the frame. You now need to fill it with your own personal story and your professional journey. I sincerely hope you took some time to jot down some key take aways, went through your own self-reflection and asked more of your own questions in the *WWHW* framework provided at the end of each chapter. I am positive that working on yourself to realign your mindset and using the powerful combination of Real Intelligence and Artificial Intelligence is the antidote for being Unblocked.

I named my consulting firm **Avigna** for a specific reason. In Sanskrit, the word means **"Remover of Obstacles."** The company slogan is *Finding Solutions, Enabling Success.*

My promise—the **Avigna Promise**—is that if you prioritize **Real Intelligence** over Artificial hype, and **Value** over Fear, there is no obstacle you cannot overcome.

Stop buying the fear. Start building the future.

To Your "Unlimited" Success,

Asim Kumar

ABOUT THE AUTHOR

Asim Kumar is a veteran Solution Architect, Business Advisor, lifelong student of the financial markets and human behavior, the founder of **Avigna**.

With over three decades of experience spanning network and security engineering, IT Strategy, financial risk management, and business development, Asim has built a career on one premise: **Real Intelligence must lead Artificial Intelligence.**

He is the creator of **The Avigna Methodology**, a framework designed to help organizations and individuals remove obstacles, align People, Process, and Technology, and move away from "Fear-Based Decision Making" towards "Unlimited Success."

Through his firm, Avigna, Asim continues to apply these principles by offering advisory, consulting, coaching and training services. Asim works with forward-thinking organizations, business leaders and individuals globally navigate the intersection of technology, financial risk and human potential.

To learn more about his work...

Connect with Asim:

- **Website:** https://www.avignallc.com

- **LinkedIn:** https://www.linkedin.com/in/asim-avigna

- **Instagram:** asim.avigna

- **X:** @asimavigna

- **Email:** asim@avignallc.com

THE AVIGNA BLUEPRINT

A Quick Reference Guide

We have covered a lot of territory in this book. From the psychology of fear to the architecture of the successful mindset to build value.

When you are in the heat of battle—whether negotiating a contract or troubleshooting a crisis, you may not have time to re-read a whole chapter. You need the codes.

Use this Blueprint as your compass.

THE CORE PHILOSOPHY

The Era of Noise: We are bombarded by threats (Security, AI, Obsolescence).

The Enemy: Fear-Based Decision Making.

The Solution: The Avigna Methodology.

The Intelligence Equation: Artificial Intelligence (Tech) + Real Intelligence (People) = UNBLOCKED

- **AI provides:** Speed, Scale, Pattern Recognition.

- **RI provides:** Context, Empathy, Ethics, Judgment.

- ***Rule****:* Never use AI to replace human judgment; use it to empower it.

THE FOUNDATION (PPT)

Success requires balancing the tripod.

- **PEOPLE (The Heart):**

    - Hire for **Willingness**, not just skills.

    - Avoid the **"Brilliant Jerk"** (High Skill, High Ego).

    - Build **Emotional Intelligence** (Psychological Safety).

- **PROCESS (The Map):**

    - Process is the removal of decision fatigue.

    - **Kaizen:** Continuous incremental

improvement.

- Avoid **Tribal Knowledge** (processes that live only in someone's head).

- **TECHNOLOGY (The Amplifier):**

  - Tech amplifies the process.

  - Automating a bad process creates chaos faster.

THE DECISION FRAMEWORKS

Selecting Solutions: P-R-O-D-U-C-T

- **P–PERFORM:** Does it do what it promises? (Integrity).

- **R–REVOLUTIONIZE:** Does it evolve the business or just maintain it?

- **O–OPTIMIZE:** Does it remove friction and save energy?

- **D–DURABLE:** Is the vendor stable? Is the solution scalable?

- **U–USABLE:** Can the humans actually use the interface?

- **C–CONSISTENT:** Is it boringly reliable?

- **T–TRANSPARENT:** Are costs and code visible?

Delivering Value: S-E-R-V-I-C-E

- **S–SUITABLE:** Is it fit for purpose?

- **E–EFFICIENT:** Does it respect the client's time?

- **R–RELIABLE:** Is it available in the storm?

- **V–VERIFIABLE:** Can you prove the value with data?

- **I–INTERACTIVE:** Is it a conversation or a transaction?

- **C–CAPABLE:** Do we have the competence to execute?

- **E–EXCELLENCE:** Is quality a habit?

Managing Economics: P-R-O-F-I-T

- **P–POSITIONING:** Are you Low Cost or Premium Value?

- **R–REVOLUTIONARY REVENUE:** Recurring value (SaaS/Managed Services).

- **O–OPERATIONAL OVERHEAD:** Cut Fat (Waste), not Muscle (Training).

- **F–FLEXIBILITY:** The ability to pivot without breaking.

- **I–INNOVATION:** Continuous improvement to protect margins.

- **T–TRANSFORMATION:** Reinvesting profit to build the future.

THE EXECUTION ENGINE

The Definition of W.O.R.K

Work is not a grind; it is an investment. "**No job is too small**."

- **W – WILLING** to...

- **O – OBTAIN**...

- **R – REAL**...

- **K – KNOWLEDGE.**

The Mindset of H.A.R.D

Hard work is not suffering; it is character.

- **H – HAPPY:** Finding joy in the puzzle.

- **A–ATTITUDE:** Being a solver, not a victim.

- **R–RIGOR:** The discipline to be thorough.

- **D–DISCIPLINE:** Consistency over intensity.

The Multiplier: S.M.A.R.T

- **S–STUDY:** Continuous learning.

- **M–MASTER:** Moving from competence to intuition.

- **A–APPLY:** Execution immediately after learning.

- **R–REVISE:** The feedback loop.

- **T–TRACK:** Measuring results.

THE MODERNIZATION METHODOLOGY (E.A.S.I.E.R)

How to adopt Modern Tech - Cloud and AI without chaos.

- **E–ELASTIC:** Infrastructure that breathes (scales up/down).

- **A–AGILE:** Velocity. Pivoting fast.

- **S–SECURE:** Reputation management. Zero Trust.

- **I–INTELLIGENT:** AI + RI.

- **E–ECONOMICAL:** FinOps. Spending based on value.

- **R–RISK REDUCER:** Resilience and redundancy.

THE STRATEGY: F.O.C.U.S

- **F–FINDING:** Hunting for signals in the noise.

- **O–OPPORTUNITIES:** Reframing problems as chances to serve.

- **C–CREATING:** Action is the antidote to fear.

- **U–UNLIMITED:** Removing self-imposed limits on potential.

- **S–SUCCESS:** The alignment of People, Process, and Tech.

Your Thoughts, Self Reflection and Action Plan

(Write how you will use the ideas presented in the blueprint and create your own experiences)